Wreck Diving

A Guide for Sportdivers

Wreck Diving

A Guide for Sportdivers

Dick Geyer

NEW CENTURY PUBLISHERS, INC.

Cover photograph by Larry Ferlan.

Printing Code
11   12   13   14   15   16

Library of Congress Cataloging in Publication Data

Geyer, Dick, 1939–
  Wreck diving.

  Includes index.
  1. Underwater exploration.   2. Shipwrecks.   I. Title.
GC65.G49   1982   627'.72   82-10575
ISBN   0-8329-0131-8

Contents

I wish to express my sincere appreciation to Beverly Geyer, Donald Groves, and Joseph Birckbichler for their generous service as divers and photography models appearing in this book.

1

Why Wreck Dive?

Why wreck dive? One of the happy illusions of wreck diving is of suddenly swimming upon a sunken Spanish galleon laden with treasures from the New World. Visions of rotting timbers from a ghostly, shattered hull, spilling out gold dubloons, pieces of eight, precious jewels and gilded swords are easily associated with the underwater world. In reality, there is little liklihood of a diver finding the treasures we all dream about. Curiosity and challenge are more appropriate reasons for wreck diving. Enjoying the opportunity to explore the rusting hulk of a once proud ship and peering into the eerie darkness of machinery locations do provide the fascination that prompts most divers to visit wrecks.

For the novice wreck diver participation need not be complicated or expensive. Standard equipment used for most other diving will usually be adequate. As a diver gains experience and as the activity becomes increasingly sophisticated, then specialized gear can be added.

Dive shops provide the best local resources for becoming involved in wreck diving, and most shops around coastal areas offer charter trips to a variety of wreck sites or can refer the interested diver to a boat captain who specializes in diving charters.

Scuba instructors are another source of valuable information. An increasing number of instructors teach wreck diving courses as part of their continuing education programs. The expert guidance provided by a professional instructor will enhance a diver's ability to participate safely in all aspects of wreck diving activity. For the vacationing diver many resorts provide excellent opportunities to visit well-known wrecks as part of their standard scuba package. Some of these wreck sites are so glamorous that they have been used as settings for major motion pictures.

Use of this text will provide the necessary information for getting started in an exciting aspect of scuba diving. Those already involved in wreck exploration will find the material useful for expanding their diving skills and the appreciation of a well-planned adventure. The format of this book is designed to support wreck diving technique. As an instruction manual, its coverage details equipment required for beginning and advanced level exploration. It also shows the reader how to determine the location of wreck sites through the use of printed sources, charter activities, navigation charts, and government agency archives.

An explanation of what to expect when diving on various wreck types will introduce the diver to the condition of steel and wooden hulled vessels after decades of lying on the sea bottom. Environmental problems the diver may encounter are thoroughly described so that adverse conditions can be anticipated. On-site diving techniques and wreck diving hazards are covered in depth to enhance the diver's ability to perform safe exploration. The taking of artifacts from a wreck is approached throughout with the recommendation that all divers leave wrecks intact so that others will have the pleasure of seeing underwater objects from the past in their natural condition.

2

Equipping the Diver

The discussion of appropriate life support equipment for safe wreck diving is broken down into several categories so that each article can be examined critically from the standpoint of usefulness in a variety of underwater situations.

Each piece of gear should be carefully chosen to enhance the wreck diver's ability to cope with a specialized and sometimes demanding underwater environment. The diver must be thoroughly familiar with the function of each item and aware of any design limitation which may compromise underwater safety.

Statistically, it can be shown that diver errors and improper diving techniques, rather than equipment malfunctions, generally contribute to diving accidents. However, when considering the inherent requirements of swimming any wreck, the diver would be wise to consider that part of proper technique includes the selection of appropriate life support gear.

3

SCUBA TANKS

As with other forms of sport diving, the single scuba tank is the most commonly used for general wreck diving situations. The 72-cubic foot steel tank and the aluminum 80-cubic foot tank provide an adequate air supply for exploration in shallow water and external excursions at moderately deep sites. The 80-cubic foot tank has become a popular choice of the diving fraternity because it can provide the diver with almost 20 cubic feet more air than the steel 72. This is because the steel 72 holds 72 cubic feet of air only when filled to 2475 psig, which is a 10% overfill. Normally the steel tank is filled to 2250 psig. At this lower pressure it holds a volume of only 64 cubic feet.

Many divers who choose to use a single tank believe that by doing so they will not get decompression sickness because the limited air supply does not provide long enough bottom time to permit them to get into trouble. This a serious error in diver judgment, since repetitive dive situations, low respiratory rates, individual diver physiology, or environmental conditions can also be factors in causing the bends even with a limited air supply.

A number of reasons can be cited for choosing a single tank; among them, it is more economic, less bulky and lighter in weight than double tank rigs, and is slightly positive-buoyant at the end of the dive. Whatever the reason, the diver should not be denied an opportunity to enjoy the experience of diving a wreck if it can be done safely on a single tank.

Double tank rigs are often used by very active wreck divers because of the wide range of sites being explored. Venturing to deeper depths, making limited penetrations of wreck hulls where safe access permits, and longer bottom times for artifact hunting all dictate the use of double tank set-ups.

Any of the above situations might be mistakenly thought to promote unsafe diving practices. However, the serious and experienced diver who goes deep or penetrates well into a wreck is generally aware of the conditions existing at a particular site and plans the dive to include the hazards involved. Diving with double tanks can also introduce a margin of safety in the event of temporary entanglement or an unplanned need to decompress. The extra air supply is comforting to have when a diver suddenly discovers, for example, fishing line snaking around the tank valve.

Double tank rigs are made up with two basic valve configurations. One utilizes a common manifold with the on–off knob situated in the center of the piping system. The other incorporates a twin bar yoke which allows two single tanks to be mounted together with independent on–off valves on each tank and a center post for attaching the scuba regulator. The advantage of using a double tank rig with independent on–off valves is that it allows the diver to swim a wreck using only one tank while the other is held in reserve for emergency situations.

Another system of double tank rigs uses no common piping and is considered by cave divers, ice divers and wreck divers to be the ultimate choice for safety in potentially hazardous environments. This double setup utilizes two single tanks with a scuba regulator attached to each cylinder. In this configuration the diver has the advantage of two completely independent air delivery systems. If one side fails or its air supply is exhausted, the diver simply switches to the other tank unit, often without missing a breath.

PONY BOTTLES

Pony bottles have become increasingly popular with wreck divers because of the additional margin of safety afforded as an independent emergency back-up system. The pony

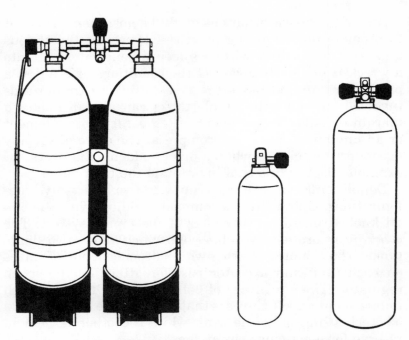

Popular scuba tank sizes: 80-cubic foot aluminum; 50-cubic foot doubles with yoke; 15 cubic foot "pony tank".

tank is constructed of steel or aluminum; with its on–off valve and scuba regulator attached, it looks like a "baby" scuba tank rig. Most pony bottles hold 15 cubic feet of air, usually more than adequate to sustain a diver in emergency situations such as being out of air, entanglement, entrapment or making an unplanned decompression stop.

Pony tanks are recognized for enhancing diver safety regardless of the type of dive being made. Especially during out-of-air situations, independent action on the diver's part will reduce anxiety because he or she is totally aware of gear placement. In one simple maneuver the diver can respond to an emergency with expedience. A diver in trouble is generally better off psychologically to trust one's own skills and equipment than to be entirely dependent upon a buddy who might not be close at hand when an emergency occurs.

The diver's choice of air supply delivery for a particular wreck dive is dependent upon the conditions known or suspected to exist on the site. Choosing the wrong tank rig could predispose the diver to additional risks.

SCUBA REGULATORS

Each piece of equipment utilized by the diver must complement the entire array of gear to provide comfort and safety while wreck diving. If, however, one item could be considered *the* most important, it would be the scuba regulator. Wreck sites and the underwater environments in which they lie can place demands upon the diver not ordinarily encountered in general depth-range diving activities.

The scuba regulator must be capable of sustaining the diver with adequate air flow at moderately deep depths. Regulators employing a balanced first-stage design are desirable for dives within the 60–100 foot range since their breathing characteristics will not be affected by depth.

Balanced type first stages are available from all scuba equipment manufacturers and may be constructed with either a flexible rubber diaphragm or metal flow-through piston in the balancing chamber. The diaphragm type is generally more sensitive to pressure changes, but because several moving parts are required for function there is greater maintenance and fine tuning associated with this model.

Heavy currents flowing at some wreck sites or increased diving depths dictate the use of a scuba regulator with a large area exhaust port in the second stage. Since exhalation resistance will promote hyperventilation and result in carbon dioxide build-up from increased expiratory effort,

the diver must be acutely aware of the regulator's ability
to exhaust easily.

It is a little-known fact within the sport community, but
well-documented by diving physiologists, that exhalation
resistance rather than inhalation effort is of the greater
importance in the respiratory cycle underwater. Especially
during moderately deep diving or increased diver effort situations, expiratory resistance becomes critical for diver
safety.

Single-hose regulator.

Another environmental condition which should be considered is cold ambient water temperature. Water or air
temperatures colder than 45°F can cause regulator freeze-
up in both stages, resulting in sudden termination of air
flow to the diver. Therefore, when diving in cold weather
or cold water situations it is advisable to have the scuba
regulator equipped with anti-freeze protection. Most diving
equipment manufacturers recognize the potential for regulator freeze-up and supply at least one model in their prod-
uct line with some modification in the first stage to protect
the unit from icing.

For ultimate safety in any wreck diving situation it is
prudent for the diver to use a scuba regulator designed with

a balanced first stage, low-temperature protection, and a large exhaust port in the second stage.

OCTOPUS UNITS

The addition of an octopus unit to the scuba regulator will enhance diver safety in out-of-air situations. If both individuals in the buddy team are wearing octopus systems they should be adequately equipped to handle air termination emergencies with confidence. Thus, since there is no need to share a common mouthpiece, confusion and anxiety can be minimized.

Octopus second stages are available with standard length (26 inches) hoses, or long hoses (36 inches). The longer hose permits the buddy team to swim in a horizontal attitude side-by-side when there is a possibility of entrapment, or assume the face-to-face position for direct verticle ascent to the surface. A disadvantage of the longer hose is the increased possibility of gear entanglement as the diver moves about the wreck site.

Several manufacturers market octopus units with a high-visibility international orange-colored second stage designed to permit the out-of-air buddy to locate the octopus with a minimum of groping in a tense situation.

If the octopus second stage chosen by the diver has no distinctive markings, it is quite easy to color-code the unit with high-visibility reflective tape. Such tape is waterproof and available at most hardware or discount stores. Being reflective, the tape enhances visual awareness even in low-light situations underwater.

Placement of the octopus unit in the diver's equipment configuration is not standardized. However, most divers generally locate this important piece of emergency life support at chest level, attached with a lanyard clip to the buoyancy control device (BCD) or tank harness shoulder strap

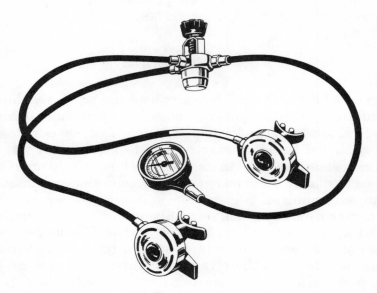

Octopus regulator.

so that the octopus can be readily seen in an out-of-air emergency.

Many authorities suggest that when a buddy-breathing situation arises, the out-of-air diver will instinctively reach for the primary regulator in use by his buddy. This seems reasonable, since it is obvious that the primary regulator is still delivering air to the buddy who is cycling bubbles. Expecting an out-of-air diver to automatically reach for the buddy's primary regulator, many divers strap the octopus unit at a convenient location, so that it is readily accessible for their own use while they give up their own primary second stage at once in order to minimize panic.

Again, it is important to remember that both divers in the buddy pair should be wearing an octopus or other emergency air source so that each has the advantage of a back-up system.

BUOYANCY CONTROL DEVICES

To the uninformed or unpracticed diver, the buoyancy control device is often considered to be primarily an inflatable life vest for surface support. However, buoyancy compensators aid the diver in a variety of ways by providing an instantaneous change in desired buoyancy. Depending on the amount of inflation or deflation of the BCD, a diver can hover motionless several feet above some part of a wreck to survey conditions prior to descent. For example, dumping air from the BCD to become negatively buoyant will allow the underwater photographer to stabilize on the wreck for accurate subject framing, and permit a diver hunting for artifacts to search the sea bottom in and around the site. Adjusting buoyancy to neutral affords the diver a streamlined swimming posture which will reduce fatigue and help conserve the air supply while navigating an area.

Surface snorkeling will also be aided if the diver remembers to achieve neutral buoyancy before starting to swim. Proper adjustment requires the diver to assume a motionless and vertical position while inflating or deflating the BCD so that the eyes are level with the water surface. Especially for long snorkel swims, correct buoyancy at the surface is desirable.

There are three basic types of BCD's worn by wreck divers. All provide the same function of buoyancy control. Some produce more buoyant lift than others, but generally each configuration type is acceptable for wreck diving situations.

The yoke, or horse collar, type has been the standard in BCD design for many years. This model has the advantage of being compact in size and close-fitting to the diver's body, making it less prone to hang up on obstructions protruding from a wreck.

Two other types, both backpack-mounted BCD's, are becoming increasingly popular with the sport diving community for all forms of activity. One is known simply as the back inflation unit; the other is a stabilizer (wrap-around) jacket unit. Both models are large-volume types having 43–53 pounds of buoyant lift and provide a degree of comfort to the diver by suspending the upper body in much the same manner a hammock does.

When backpack type BCD's were first introduced to the market there was some resistance by wreck divers to the use of this configuration because it was believed the large bag would not permit access into confined spaces, and that the bag could become fouled on obstructions unnoticed by the diver. However, the inflated bag's dimensions are not much wider than the diver's body and its height, when horizontal, does not present enough area to prohibit access to small areas.

All top-quality BCD's are constructed of a tough and abrasive-resistant outer bag material which should preclude the possibility of tearing or cutting on wreckage, and perhaps the most important questions to consider in BCD selection for wreck diving should be: does the BCD provide a minimum of 35 pounds of buoyant lift, and will it automatically float an unconscious diver on the surface in a vertical or face-up position to prevent drowning?

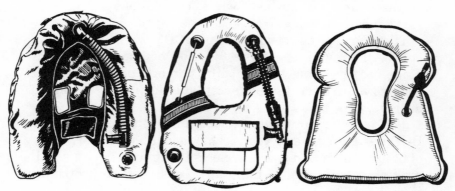

Buoyancy control devices.

INSTRUMENTS

Submersible Pressure Gauges

The use of a submersible pressure gauge (SPG) for continuous monitoring of the diver's air supply is mandatory in all wreck diving situations. SPG's complement safety since these units provide valuable information on air consumption and when a dive should be terminated to allow a normal controlled ascent. Monitoring the SPG will show the diver whether he has enough air remaining for decompression stops, planned or unplanned.

Divers who anticipate making penetrations into wrecks will usually plan the dive based on the "rule of thirds" for air consumption: use one-third of the air supply descending and penetrating; one-third for retreat and ascent; the remaining one-third is held for emergencies. This is a sensible rule to follow because the margin of safety introduced should

Console of depth gauge
and submersible pres-
sure gauge.

prevent a diver from suddenly running out of air in the
course of the dive. The submersible pressure gauge should
be taken very seriously. It is the most valuable instrument
in the life support system.

Watches

For all wreck diving activities some sort of timing device
is mandatory in monitoring bottom time. Few wreck sites
are found in depths shallower than 33 feet, therefore most
divers make repetitive dives on the sites being worked.

At least one member of the buddy pair must wear a timing
device to ensure that every dive is made well within established no-decompression time limits. In fact, the entire dive
plan should be established to set maximum bottom times
for each dive and an adequate surface interval between
dives to prevent the need for staged decompression or bends.
Consider the possibility of a buddy team using double tank
rigs on a repetitive dive to 60 feet, exceeding the no-decompression time limit: without the use of a timing device,
how would the divers know of the error, and how would the
resulting required decompression time be arrived at? Especially on repetitive dive situations, it is quite easy to exceed no-decompression time limits. Therefore, accurate down
time must be recorded to promote safety.

Diver's wrist watches are available in a variety of styles
and movements. Some are of better design than others, and
design features are important considerations in selecting a
watch for wreck diving. A good dive watch should be pressure-proof (not the same as water-proof) to a minimum depth
of 330 feet (10 atmospheres) to be considered adequate. The
movement can be either automatic self-winding or quartz.
Watches that must be manually wound each day are not
considered good time pieces because continued winding of
the watch stem wears out the seal between the stem and

case. This very quickly increases the possibility of flooding the watch while underwater. A large-faced, highly luminescent dial with one-way rotating ratchet bezel will enhance the diver's ability to monitor bottom time during the dive. All of these features indicate a moderately expensive diver's watch. However, the diver's safety dictates that an accurate and ruggedly built time piece be used for wreck diving.

BOTTOM TIMER

A very popular timing device, universally accepted for all types of diving, is the bottom timer. This unit is a simple stopwatch that has been encapsulated in a pressure-proof plastic case. Known by the trade name Bottom-Timer, it incorporates all of the desirable features required for accurate timing. It has a large faced, highly luminescent dial, sweep second hand and long minute hand which is pressure activated by a minimum water depth of 5–7 feet. The device is direct-reading from 0–60 minutes and will continue to run as long as the diver stays deeper than the minimum pressure activating depth. Between dives, when the minute hand is reset to zero, or for periodic winding (timer runs 30 hours on one winding) are the only occasions when the unit needs attention.

DEPTH GAUGES

Water depths at wrecks vary from a few feet at on-shore sites to more than one hundred feet at sites in the open ocean or steeply contoured bottoms, such as those found in the Great Lakes. Because of the wide range of depths, especially for depths in excess of 33 feet, it is mandatory that all wreck divers wear an accurate depth gauge.

There are currently three basic types of depth gauges produced for diving: liquid filled (silicone liquid), capillary and open bourdon tube. Of the three types the liquid filled is the most accurate through its entire depth range. Manufacturers advertise an accuracy of two feet plus or minus at any depth on the gauge scale. Since the mechanism is sealed in a liquid-filled housing, depth sensing is very precise because the liquid transmits pressure to the gauge. These units are also relatively rugged because the liquid absorbs minor shocks that could damage nonliquid-filled mechanisms, and because the pressure-sensing tube is never exposed to the corrosive action of water.

Capillary depth gauges are simple in construction and incorporate an open-ended hollow tube that allows water to pass and pressurize air trapped inside the tube. This design works on Boyle's Law, the principle that gas volume changes as pressure on it increases or decreases with depth changes. With its associated foot-marking scale the capillary gauge is extremely accurate at shallow depths, usually from zero feet to 40 feet. Deeper than 40 feet, this type gauge is not reliable because the spaces between depth scale markings are dramatically reduced and therefore difficult to read. Due to its accuracy at shallow depths, wreck divers use the capillary gauge in conjunction with an oil-filled unit in anticipation of needing to make precise depth readings for decompression stops.

The diver should be aware that the capillary tube opening can become clogged with small grains of sand, silt or algae. Even partial occlusion of the tube can cause separation of the water column and reduce accuracy. Periodic removal of the tube and swabbing with a pipe cleaner to clear the tube of obstructions will enhance the reliability of the capillary gauge.

By design, open bourdon tube depth gauges are not as accurate in depth-sensing as are liquid-filled units. This type of gauge incorporates either a flexible rubber diaphragm or a small screen fitted over a copper alloy device called a "C" tube. Water enters the gauge housing through an open port and creates pressure on the "C" tube which causes the tube to flex as the diver descends or ascends, thus moving a needle pointer calibrated to indicate feet of depth.

Since this type unit senses pressure through an open port in the gauge housing, some parts are exposed to sea water. Continual submerging and drying between dives causes salt and mineral deposits to accumulate in the porting unit. After a period of time the port becomes too encrusted for water pressure to activate the mechanism. A thorough freshwater rinse after each dive will help reduce deposit material. In addition, it is suggested that after every fifth dive the gauge be completely submerged in a glass of distilled white vinegar for 30 minutes, then rinsed thoroughly with a gentle flow of water into the port area. Gently shaking the gauge will remove water droplets from the port and promote drying.

Depth gauges are essential for safety in the prevention of decompression sickness and to provide data for future reference in planning a dive at a particular wreck site.

COMPASS

There is a basic need for navigation on almost every dive. Occasionally it is possible in well-known areas, with good visibility, to use bottom topography, angle of sunlight or other environmental cues for simple navigation in a wreck site area. However, diving conditions are rarely favorable for noncompass swims.

The experienced wreck diver understands that a compass is a valuable instrument which affords straight line navigation to and from the site. Air consumption and total dive time can be calculated based on compass swims. Also, accurate plotting of the wreck location for future dives will be aided by recording bearings of the area in degrees. Since all diver compasses utilize a magnetized needle that aligns with flux lines in the earth's surface (pointing toward the magnetic north pole), selection is usually based on cosmetic features rather than on accuracy.

DECOMPRESSION METERS

At present, there are two "decompression meters" on the market for aiding divers in the prevention of decompression sickness when making deep and/or repetitive dives. Neither unit should be relied upon absolutely to prevent bends. Prudent use of decompression meters includes monitoring time and depth, plus strict observance of the United States Navy decompression tables.

All mechanical or electronic instruments are subject to failure and diver error. Therefore, the diver's health and welfare, with respect to decompression sickness, must never be dependent upon metered information. Individual diver physiology is too complex for meters to accurately correlate dive profile data with nitrogen absorption in the diver's body.

Wreck divers anticipating the use of a decompression meter are advised to become thoroughly familiar with the meter's design limitations and how the unit functions prior to any rigorous diving activities.

SAFETY EQUIPMENT

Diver Knife

A sharp knife is an absolute necessity for diver safety when swimming around wrecks. There are many vivid tales, mostly true, about divers needing to extricate themselves or a buddy from the web-like entanglement of fishing line, trawler net, or hull rigging.

Wreck sites provide an artificial habitat for a variety of marine life forms in saltwater and freshwater environments. Fishermen know these areas as hot-spots providing abundant catches with line or net. Many wrecks are initially discovered through fishing, because as word about a good, deep fishing spot spreads, divers often will investigate the area and find a sunken vessel. Imagine the amount of fishing line or net that will be snagged and abandoned on hull wreckage after months, even years, of fishermen working such a site. Lazily trailing in a gentle current, these

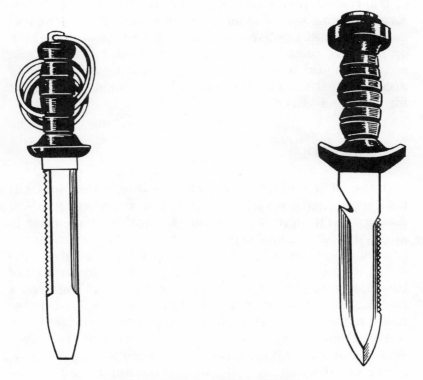

Diver's prying tool and knife.

unseen hazards can entangle the diver in a frightening embrace. Occasionally, a cool-thinking diver will be able to slowly work free of line entanglement. However, a sharp knife does the job quickly, completely. The diver and buddy must be careful while cutting lines that some part of the anatomy or gear of either is not slashed in the excitement of trying to get free.

Many wreck divers wear a small stiletto type of knife on either arm as a back-up to the regularly used cutting tool so that at least one knife is always accessible. Most small knives are designed with extremely sharp edges, razor or serrated, which can slice through line without sawing and swearing.

The classic diver's tool with its blunt-nosed blade is perhaps the most versatile for working about wrecks. The tool can be used as a pry bar when the diver needs one in working on a superstructure or bottom debris. Probing sand areas, scraping metal to reveal brass, and cutting a line can be accomplished with a diver's tool. It is a handy implement for the wreck diver's kit.

Underwater Lights

In the basic scuba course divers are taught that sunlight does not penetrate very far below the water surface. The passage of sunlight through any natural body of water is affected by suspended organic and inorganic particles, the time of day, type of bottom and other environmental variables. Because water acts as a filter, selective absorption of the color spectrum quickly becomes apparent. The diver's inability to detect vibrant, pastel tones even a few feet beneath the surface creates a distorted view of the underwater world. Even during a daylight wreck excursion, the diver should make a habit of carrying an underwater light. Using artificial illumination will enhance the ability to perceive striking colors and can aid visibility, especially when peering under ledges or into otherwise ink-black holes.

Underwater lights are available in a variety of shapes, sizes, beam brightness and power sources. There is available such a wide selection that many divers find choosing a light frustrating. For general wreck diving almost any underwater light will provide adequate illumination for inquisitive observation. Smaller sized units serve this purpose nicely.

Divers anticipating even limited wreck penetrations should consider using lights with powerful beams and possibly a second small light as a back-up in the event of primary source failure. Maximum illumination is required for penetrations because there will be a total absence of ambient sunlight inside a hull wreckage. A fresh battery or fully charged "ni-cad" pack should insure beam output during the entire dive.

Waterproof Decompression Tables

No diver should consider making even a relatively shallow wreck dive without taking a set of decompression tables along underwater. There is always a possibility of exceeding no-decompression limits due to low respiratory rates, entanglement, loosing track of time or having dived deeper than planned. The need for unplanned decompression can occur suddenly during a dive. Therefore, having tables at hand will provide the means for figuring appropriate stop depths and times to help prevent decompression sickness.

Correct application of the decompression tables is essential. There is no safety factor built into the tables nor can there be even the slightest error in figuring schedules. Frequent practice in decompression table usage will be rewarded should a real need for it suddenly develop.

Slate

There is perhaps nothing more frustrating on a dive than the inability to communicate. Several years ago, a set of hand signals were recognized internationally as a means of simple communication. Although hand signals graphically convey a few basic directions between buddies, they have limited application for most wreck diving situations.

Diver slates provide a positive means to write out messages with little chance of misunderstanding. With them, for example, wreck hunters can plot locations for future investigation, photographers can direct models into staged positions or record exposure details, and buddies can indicate changes in the dive plan or figure decompression requirements. The diver's slate is a versatile item. Its potential should not be ignored.

Safety Line Reels

Penetrations into wreckage require emotional stability, alertness and the proper equipment. Finding one's way through a tangled maze of collapsed superstructure and back again is an impossible task without the aid of a safety line. They can be made of almost any type of rope, from thin clothesline to high-strength nylon line wound on a special reel housing.

The reel system is a unique, compact unit easily handled in dark and confined spaces. As the dive team prepares to penetrate, a snap at the end of the line is secured to some sturdy piece of outside wreckage. Entry is begun with the lead diver letting out line while cautiously keeping tension on it to prevent snagging. The buddy diver maintains hand contact with the line at all times to prevent becoming disoriented. Upon reaching the planned penetration distance, the dive team turns back, reversing the original course while reeling in the line.

At present, safety line reels are not commercially marketed. However, a sturdy system can be assembled from spear gun materials available at most dive shops.

THERMAL PROTECTION

Wet Suits

The selection of appropriate thermal body protection for wreck diving is based on expected water temperature of the area. Obviously, the type of suit used for diving wrecks off the New Jersey coast would not be desirable for warmer Caribbean waters. In tropical areas a full suit made of one-eighth inch thick neoprene material is adequate. It will insulate the diver sufficiently and at the same time protect against cuts or abrasions on body extremities during movement around the site.

For diving all northern waters maximum thermal protection is required. Thermoclines are encountered in most areas with the resultant temperature change being as much as 20 degrees colder under the thermocline. Deep diving causes wet suit compression and loss of thermal protection due to the increase in surrounding water pressure.

Jackets without arm zippers, "Farmer John" style pants without leg zippers, cold water hoods (bibbed), three-finger mittens, all constructed of one-quarter inch thickness neoprene, are usually the choice of northern divers to provide good insulation against cold water environments.

Many divers fit into a standard size offered by wet suit manufacturers, some do not. Trying a suit on at a local dive shop is the only method for making an intelligent decision with regard to proper fit. For maximum protection the wet suit must fit snuggly, but comfortably, in all body areas.

Farmer Johns.

There cannot be air pocket bulges, loose fitting wrist or ankle circumferences, or length variations. Any of these excesses will cause continual flushing of water in and out of the suit. Trapping a minimum film of water inside the wet suit is the secret of good thermal protection. If there is doubt concerning good fit, the diver would be wise to invest in a tailored suit cut to exact body dimensions. The comfort provided by a custom wet suit will offset the small additional expense.

Dry Suits

The "ultimate" in thermal protection best describes a dry suit. Regardless of water temperature or depth, even in howling winds on deck, the diver stays warm because a layer of air surrounds the body.

Dry suit features include attached hood with neck seal, attached boots and tight-fitting wrist seals, all designed to prevent water from entering. The somewhat bulky dry suit construction permits the wearing of clothing or underwear that increases insulation from cold environments. Initial dives with a dry suit should be made only after thorough instruction and open water check-out from a competent individual. This process is necessary so that the diver is made aware of how air volume changes within the suit affect routine diving techniques.

Dry suits are expensive and could be considered a luxury. However, for deep diving and/or extremely cold water situations the dry suit provides an additional margin of safety which should be considered.

Weight Belts

Years ago, divers adjusted weight belts according to the depth of a dive. For shallow dives neutral buoyancy on the surface was achieved so that exhalation would cause the diver to descend slowly. When preparing for deeper dives it was not unusual to eliminate the weight belt entirely since wet suit compression was sufficient to keep the diver at or near neutral buoyancy while swimming at depth. Thus it can be a struggle to get down the first 10 feet, but once on the bottom trim is close to ideal.

Over the years, and especially since the introduction of large-volume buoyancy compensators, divers have lost the "art of proper weighting" for a particular type of dive. In fact, overweighting has become a problem with today's divers. So much so that heavy weight belts have been implicated as a contributory factor in many of the drowning accidents involving scuba divers. It is not unusual to hear a charter boat captain question divers about excess lead on the weight belt when a final equipment check is conducted before water entry.

To ensure personal safety, only enough lead should be used to off-set the positive buoyancy gained by wearing a wet suit. Properly weighted, the diver should be able to float vertically at eye level on the surface with *no* air in the BCD and holding only a normal breath.

Although nylon webbed belts are customarily used, rubber compensating belts provide a more secure method of weight attachment. Nylon belts have a tendency to become loose and slide around or below the waist as the diver descends. This is caused by wet suit compression as water pressure increases. Since the rubber belt is elastic, it will conform snuggly to the body with changes in depth. Compensating rubber weight belts are especially desirable for divers wearing dry suits because air is continually being introduced and exhausted to maintain body warmth or correct for buoyancy. The rubber belt will permit suit volume changes without becoming loose.

Accessories

Many accessories are used by wreck divers to augment a particular dive. They should be selected with emphasis on utility and their compatibility with safety.

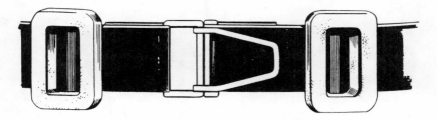

Weight belt with wire quick-release buckle.

Nylon mesh bags are considered standard gear for wreck diving. They serve so many functions the diver would feel naked without one. Holding captured lobsters, small tools or recovered underwater items are only a few ways these bags can be useful.

Marker buoys, pipe wrenches, hack saws, crow bars and lift bags enhance the wreck diver's ability to salvage artifacts from a site.

Underwater photographic equipment allows the diver to fully document an excursion for personal reflection of a memorable diving trip. Data collection regarding artifact salvage or historic preservation will be aided with pictorial records.

When considering accessories, keep in mind that excess task loading is produced by trying to achieve too many objectives during a single dive. If photography on the site is planned, there is no reason to carry a bag full of salvage tools also. Most divers are inclined to overload their accessory needs when preparing to dive a wreck. This temptation must be resisted. Carry only those items necessary to accomplish the planned objective.

3

Wreck Hunting Technique

It is very rare for divers to suddenly encounter a wreck site while swimming a favorite area without having some prior knowledge of the existance of the wreck. While there are thousands of sunken hulls lying on the bottom of coastal areas and inland navigable lakes, unexpected finds are almost beyond the bounds of reality. Locating wreck sites can at times be a demanding and lengthy research process, though often, discovery is no more detailed than simply communicating with other divers. In anycase, the quest always begins on shore.

Initially, the diver is faced with a seemingly insuperable challenge of locating a clue. Some areas, like the New Jersey coast, are recognized as virtual graveyards for sunken ships, while other coastal regions and many inland lakes have been ignored as potential sites for exciting wreck exploration.

Local sources can provide valuable details about the types of wrecks to be found in a region. If area waters encompass established shipping lanes or approaches to major harbor facilities, a diver could expect to find in them a variety of hull shapes and sizes. Every class of vessel that floats, passenger liner to small craft, is a candidate for sinking in heavily traveled regions. Over the years a multitude of disasters will occur, and each unfortunate sinking provides the wreck diving enthusiast with an opportunity for a memorable adventure.

Inland lakes and rivers are sometimes not recognized as potential waterways that contain shipwrecks. However, the Great Lakes have a 200-year history of commercial shipping and fishing, and there are numerous wrecks on the shoals of each lake. Many are well-documented and dived regularly.

Often large navigable rivers have shipping lanes that carry a staggering amount of vessel traffic. For example, Monongahela, Allegheny, and Ohio rivers flowing through Pittsburgh, Pennsylvania provide the means to move more gross commercial tonnage than does the Panama Canal. River wrecks usually are not glamorous, but they can be an interesting source for diving adventure.

The avid wreck diver is first an information seeker who solicits knowledge from a variety of sources, and commonly, one source will lead to another so that fact can be separated from lore. Questioning other divers and local dive shop personnel is also an appropriate way to gain information about area wrecks. In-depth detail will usually be given quite freely, so that both of these sources will often provide extensive descriptions of known sites.

Accessibility to the site, expected depth of the wreck, hazards to be avoided, and environmental conditions are some of the specific questions that should be answered prior to generating a dive plan. An inquisitive diver should be capable of piecing together enough particulars to form at least a basic knowledge of what conditions to expect at a wreck site.

INFORMATION RESOURCE BOOKS

Numerous resource books are available at public libraries and bookstores that contain fasinating details about shipwreck disasters. Resource publications listing shipwrecks are not generally well known, but, because of the importance of descriptive material, an effort should be made to review as many references as are obtainable. There are at least four directory-format texts available listing shipwrecks in United States waters. Three of these cover both coastal and inland waters.

A Guide to Sunken Ships in American Waters, by LCDR Adrien L. Lonsdale/H.R. Kaplan, Compass Publications, Inc., Arlington, Virginia, lists over 1,100 major shipwrecks taken mostly from official United States Government sources. Information is provided by geographical area in easy to read tables. Each table shows latitude/longitude, type of ship, name, tonnage, depth of water, nautical chart number, and brief remarks about cargo value.

The *Encyclopedia of American Shipwrecks*, by Bruce D. Berman, The Mariners Press Inc., Boston, Massachusetts, documents disasters extending from the pre-Revolutionary War era to the 1970s. Listed aphabetically and geographically are 13,000 vessel sinkings. Data supplied include the vessel name, hull type, tonnage, year built, date of sinking, cause of disaster and land-oriented bearings. Regional maps are displayed before each section to aid the reader in area location.

A *Shipwreck Reference Report for all U.S. Waters* is available from the National Archives and Records Service, General Services Administration, Washington, D.C. This government source reference traces all major vessel strandings and sinkings in United States territorial waters in much the same format as the previously listed texts. However, the Archives report is numerically more comprehensive.

The Great Lakes area is seperately covered in the book *Directory of Shipwrecks of the Great Lakes*, by Karl E. Heden, Bruce Humphries Publishers, Boston, Massachusetts. Approximately 1,500 wrecks are described in alphabetical order for each lake and their connecting waterways.

NEWSPAPERS

Local newspapers can be important information resources rarely considered when attempting to locate wrecks. First-hand accounts of area shipping disasters may sometimes pinpoint a potential site without further editorial investigation. Even small-town newspapers are able to supply complete microfilmed records of every edition printed during their circulation history. Newspaper articles generally provide more detailed guidance than text references because of on-the-scene reporting. However, a diver researching newspaper articles should expect a degree of sensationalism to be included in any shipwreck account.

NAUTICAL CHARTS

Nautical charts and certain related publications indicate exact locations of wrecks adjacent to coastal areas. Wrecks are shown because of a possible hazard to safe navigation. Divers can take advantage of charted sites without consulting other written sources since accurate positions have been plotted. Nautical charts are available from the Department of Commerce, National Ocean Survey, Riverdale, Maryland.

Several classifications of charts exist that may be of interest to wreck divers. Each chart contains various notes, symbols or abbreviations used to identify specific information. These notations are essential for complete interpretation of the chart and should be thoroughly understood by the user. Various scales are employed to portray features in significant detail for each chart class. Hydrographic data are not generally shown for inshore areas on small-scale projections commonly known as sailing charts. Wreck divers considering the use of charts should always obtain the largest scale coverage available for a particular locale.

Small craft charts are specifically designed for inshore utilization. They are printed in large scale for easy co-ordination with land references. Small craft charts clearly mark intercoastal waterways, wrecks, obstructions, and other features of interest.

Coast charts project medium-scale design intended for coastal navigation inside offshore shoals and reefs, entrance bays or harbors of considerable size and certain inland waterways. Coast charts cover a larger area than small craft types, but are equally useful for wreck location.

Navigation charts are available for each of the Great Lakes. Projections may be obtained in local and entire single-lake coverage from the U.S. Army Corps of Engineers, Lake Survey, 630 Federal Building, Detroit, Michigan or the Canadian Hydrographic Service, Chart Distribution Office, Survey and Mapping Building, Ottawa, Ontario, Canada. Charts of the Great Lakes are not as detailed as those covering coastal navigation; however, some inshore wreck sites are shown.

Divers anticipating chart usage should realize that charts depict only wreck position, type of bottom in the area and depth of water over the site at low tide. For specific descriptive data, the diver will need to consult other reference sources.

NOTICE TO MARINERS

Notices to mariners may be obtained from the U.S. Coast Guard, Notice to Mariners, Department of Transportation, Washington, D.C. They are issued on a weekly basis with the intent of advising all interested parties of local changes in aids to navigation, including vessel strandings and sinkings, which may be considered hazards to ship traffic. These notices are issued without charge, and cover, by specific area, all coastal waters, the Great Lakes and major rivers of the United States.

Weekly advisories can be of great interest to wreck divers because recent sinkings make possible virgin wreck exploration because such vessels may be found intact. Rigging, machinery, port holes or other artifacts will be discovered in good condition, but they should not be removed.

Cautious planning is advised before diving on recent wrecks reported by notices to mariners. Legal rights protecting the vessel owner's property may be in force and should not be ignored. An understanding of prevailing maritime law will help prevent involvement in legal confrontations.

CHARTER BOATS

For many divers, the only opportunity for participation in wreck diving is by taking organized charter trips to known sites. An increasing number of boat captains are catering only to divers. Their vessels employ special electronic equipment for pinpoint location and rigging adapted to meet diver needs. Experienced boat captains can often provide complete details of each site visited. Many captains are also experienced divers and have first-hand knowledge of hull structure, profile position, expected environmental conditions and existing hazards found on a particular wreck.

Charter trips can be exciting activities pursued by novice and experienced wreck divers alike. Sharing descriptions of an eerie hulk, agonizing over the ten-pound lobster that got away, or reflecting on the discovery of a ship's brass bell add to the camaraderie of a memorable group charter.

Diver with handheld metal detector sensitive enough to detect pen-nysize objects under four inches of bottom cover and down to four feet or more with a larger masses.

Boat-towed metal detector used for attempting to locate known but uncharted sites or for searching areas suspected to have wreckage.

UNDERWATER METAL LOCATORS

Utilization of an underwater metal locator provides a new dimension in wreck hunting. Several models are available so that selection can be made dependent upon expected use or economics. Handheld detectors, exclusively designed for divers, have been on the market since the early 1960s. Still, few divers are aware of their existance and many have not considered the potential they afford to increase a diver's chances of finding important artifacts.

Artifact hunting while swimming over a well-known wreck is superficial at best. Most sites will have been wantonly stripped long ago of obvious articles valued by wreck divers. The serious artifact collector understands this and will search under the sand around a wreck for hidden objects of worth. Fanning the sand, as the technique is called, is occasionally productive, but still many artifacts will be missed.

Using underwater metal locators provides a positive means of revealing the presence of buried objects. Employing an inductance coil, sensitive meter and audio impulses, these locators will detect all metal-bearing minerals as well as ferrous and nonferous metals. The sensitivity of most handheld units will permit detection of penny-size metal objects under four inches of bottom cover and down to four feet or more with larger masses of metal. Most handheld detectors are economically priced and provide the diver with an extremely reliable instrument, ruggedly built to withstand continuous use.

Boat-towed metal detectors are also available for those divers attempting to locate known but uncharted sites or for searching areas suspected to have wreckage. Towed models are designed to be used from a boat to detect large masses of metal such as hulls, anchors, cannon, or machinery. However, these units utilize even more sensitive inductance coils than handheld models and will detect small objects as well.

4

What to Expect on a Wreck

Divers who have never visited an underwater wreck site often expect to see a sunken ship with workable machinery, intact rigging and slightly listing hull lying majestically on the bottom. A few wrecks can be viewed as intact, but they represent the rare exception. Intact wrecks generally become famous for their alluring appearance and therefore provide a popular and valuable resource for a variety of diving activities.

First-time wreck dives can also be an emotionally disappointing experience. Over many years of submergence, environmental forces may act to obliterate all recognizable traces of original structural contour and shape. Corrosion and marine growth will reduce well-defined articles of machinery to a singularly indistinct form. Decks collapse and hull supports separate, rendering a once proud vessel to a tangled maze of rubble.

The extent of corrosion and decay is dependent on several

conditions. How the disaster occurred is a major factor in hull intactness following sinking. If there was fire aboard, major structures, rigging and other articles may have been destroyed. Wooden vessels often burn to the water line, leaving little trace of deck construction. Explosions on board will devastate large hull sections. Reef strandings cause tearing and pounding through keel supports, sometimes resulting in complete separation of bow and stern as the vessel slides into deeper water.

Ocean wrecks show greater destruction from environmental forces than those found in freshwater. Articles of metal fabrication corrode quickly in saltwater. Wooden structures are attacked by marine organisms and take on a worm-eaten crumbling appearance within months of sinking.

Freshwater wrecks are not as rapidly destroyed. Metals do corrode, but even after many years of submergence moveable objects may still function and are easily recognized. Wooden articles display remarkable preservation. In fact, most woods can be brought to the surface, dried and easily restored without sign of deterioration. Many Great Lakes wreck divers have fabricated stunning pieces of furniture from wood recovered after a century of being underwater.

Though there may be extensive deterioration, each wreck still holds many mysteries and ample enticement for investigation. Photographic possibilities abound. Lobstering and spearfishing bring the game hunter. And, of course, basic recreational exploration is the attraction that initially lures all divers to a wreck site.

FREIGHTERS

There are various classes of freighters, generally typed according to hull size or gross tonnage capacity. Coasters, lighters, and container ships all serve a common purpose,

Diver viewing remains of a wooden wreck.

Inspecting the hull of a metal wreck.

Some underwater shipwrecks look like junk yards.

the movement of cargo between ports. Sunken freighters often obtain a staggering array of articles. Cargo lashed on deck before the sinking may still be located in original position. Spilled cargo will slowly settle around the site and be found adjacent to the hull. Cargo holds, if safely penetrable, are virtual junk yards, littered with goods of every description.

PASSENGER SHIPS

Passenger ships represent a more glamorous type of wreck because expensive polished accents are used in contruction of accommodation areas. There is usually an abundance of unpainted brass poles, railings, stanchions and fittings displayed on upper deck sections.

Crystal, china, silver settings, elaborate furnishings, all

used for the comfort of ocean travelers, may be discovered undisturbed. Certainly the glitter will have been long removed through years of submergence, but a little rubbing or scraping by an inquisitive diver should help reveal articles of worth.

Such popular items as ship's bells, navigation instruments, ship's wheels, engine telegraphs, and running lights are usually to be found intact, a pleasure to be shared by all divers. Many items on board a passenger ship will be engraved with its registered name.

TUG BOATS

Tug boat sinkings are generally the result of fire or flooding. Because of this, many tugs will be discovered mostly intact on the bottom. Due to the nature of the work these vessels have been designed for, unusual pieces of machinery and rigging are strategically positioned on deck. Large capstans sit fore and aft to control heavy lines when maneuvering tows. Heavy rigged booms, winches, block and tackle gear clutter the stern. A simple, but practical, wheelhouse sits just forward of amidship, allowing the pilot an unobstructed view of the surroundings.

Encountering a sunken tug boat in good visibility water is an electrifying experience. These small-sized work vessels display the most picturesque lines of all wrecks, and the photographic potential of an intact tug is enormous.

SMALL CRAFT

Small craft sinkings occur more frequently than large work vessels, generally because of improper navigation techniques, poor hull and engine maintenance, or venturing out in seas too heavy for a vessel's size. Hundreds of small craft go down every year in coastal areas, rivers and lakes. Unfortunate as this is for the boat owner, these sinkings can provide excellent opportunities to dive virgin wrecks.

Though diving on small craft may not be as intriguing as exploring large vessels, many still will have been equipped with not only standard operating gear, but also with elegant trappings used to dress up the vessel.

A shiny chromed anchor, deck cleats, ship's wheel, running lights, small portholes, rigging blocks and tackle, and propellers are highly prized articles for inspection.

As in the case of large vessels, it must be remembered that local, state or federal laws may be in force to protect the boat owner's property. Even though a sunken small craft is apparently abandoned, thorough investigation of salvage rights is advised before attempting to remove any articles from it.

BALLAST STONE

However remote the possibility may be, it is every diver's dream to find a sunken galleon. Gold dubloons, pieces of eight, gilded swords, chests filled with precious jewels, all the treasures that romantic thoughts are made of!

Finding the intact remains of a wooden hulled galleon is not physically possible, since the marine environment will have destroyed or completely obliterated all traces of the vessel's structure centuries ago. However, what may be found, which would suggest the underwater site of a sunken galleon, are ballast stones. Most galleon-era ships used stones placed inside the hull and below decks to keep the vessel on an even keel and to increase its draft so that the ship would sit low in the water for increased maneuverability.

Ballast stones were cut in graduated sizes from large blocks to relatively small pieces, thus wedging the ballast together to prevent shifting in heavy seas. Stones nearest the hull sides were generally fitted to shape and had a uniform pattern.

Seeing a pile of stones, not natural to the area, on a sandy or coral bottom might strongly suggest the site of a centuries-old wreck. Unless the hull had been badly torn open during the sinking, the ballast stones might still be in their original position relative to hull shape, although they would have spread out as the hull planking and ribbing eroded away. Any pile of rocks showing a distinctive pattern on the bottom should be investigated.

ENVIRONMENTAL CONDITIONS

As with all other locales used by sport divers, environmental conditions at a wreck site will vary, and because of potential dangers from the hulk itself, adverse conditions may compromise the divers' safety.

Poor visibility, resulting from plankton blooms, sand or silt suspension, and thermocline haze will reduce the diver's ability to detect perilous situations. Wreck divers would be wise to remember that every site has a potential for silting over while the dive is in progress. Keep alert for sudden zero visibility and react positively by moving slowly, but promptly, out of the silted area. Tides or currents sweeping through a wreck site also can create other problems. For example, visibility will be affected by silt carried in the flow. On the other hand, the diver may occasionally use a moderate current advantageously, by moving into the flow, thus having optimum visibility in the direction of travel.

Diver stability on a wreck is extremely important for work and safety. Maneuvering in heavy currents or tidal surges will exhaust even the strongest diver in relatively short time. Continual assessment of individual fatigue must become routine.

MARINE LIFE

Generally a wide variety of marine animals is to be seen on a wreck site, for a shattered or crumbling hull can provide sanctuary, an established food-chain, and a nesting area for a large population of marine animals. These creatures enhance the wreck diver's visit with animated displays of constant activity.

It has long been recognized that marine animals are defensive by nature and will usually move away from a diver to avoid any contact. However, it may not always be convenient for an animal to retreat from the groping hands and thrusting arms of an overly enthusiastic wreck diver. A denizen's sudden flight reaction or defense of territory with flashing teeth may jolt the diver into realizing why black holes and crevices should not be penetrated indescriminately. Always investigate an opening cautiously with a beam of light and probe before reaching into it.

The possibility of encountering some form of dangerous marine animal on a site must be considered, but the diver should not allow this possibility to create a mood of apprehension that will interfere with the enjoyment of investigating a wreck.

5

Wreck Dive Technique

A thoughtful dive plan developed through dialogue between buddies or group charter personnel will reduce the environmental vulnerability of the dive team and help to establish correct procedure. Wreck diving is rigorous and at times even dangerous. Therefore it is of the utmost importance for reasons of safety to develop good wreck diving techniques that take into consideration every aspect of a proposed excursion.

BOAT ETIQUETTE AND SAFETY

Most wreck diving enthusiasts participate in the sport through organized charter boat activities and for good reason. Charter boat captains know exactly where to find wreck sites offshore. Employing sophisticated electronic aids allows them to pinpoint locations over a site so that they can deftly drop anchor directly at the wreck. Locating an enticing wreck is easy for the knowledgeable charter captain, but the functions of tending a group of divers can be a complex and demanding task.

Any diver who plans to pursue charter activities must develop an awareness of boat etiquette, plus an appreciation of a captain's responsibility to his ship and his passengers.

Learning proper boat terminology is a first step toward a satisfactory familiarity with charter etiquette. For most new wreck divers, terms used to identify vessel parts are quite foreign. One need not be a bonafide sailor to understand nautical labels, but having an active interest in basic identification instills a sense of pride.

A typical charter diving boat can be thought of in terms of a floating building used to transport and house divers in support of their activities. For instance, there are stairs, rooms, floors, a front, back and center area, an entry–exit access, a system for driving the vessel and electronic equipment for directional guidance. While all of these terms are descriptive, they are not used afloat.

Looking from back to front of a boat, the back is called the stern; directly outside the stern is a platform and boarding ladder used for safe access when getting in or out of the water; the main walking floor is the deck, which is attached to the high side walls, or gunwhales (gunnels); fixed on the gunwhales are cleats used for tying ropes between the boat and dock when the vessel is moored in its berth or dock space; the cabin sits amidship, or near the center, of the deck and houses the steering console, engine controls, electronic devices, and a marine head, or bathroom. Forward of the cabin is the bow, or front area, and there will usually be a post, situated immediately behind the foremost point of the bow, that is used for tying off the anchor line once the anchor has been set at the wreck site. All of these items are referred to as being topside; spaces underneath the main deck are known as below deck areas. Below deck compart-

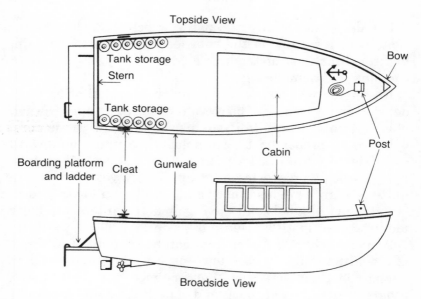

Dive boat configuration.

ments are usually restricted to captain and crew usage. Therefore, it is prudent to ask permission from the captain prior to venturing down any ladder or stairway that leads below decks.

The above diagram shows a typical charter boat's configuration. Also, a listing of nautical terms is included at the end of this book.

Proper etiquette for all divers includes thoughtful consideration and respect for another's property. Always keep in mind that when chartering a boat you are only paying for a ride and not taking over the vessel's ownership for the day.

A brief outline of charter etiquette follows as a starting point for ensuring a rewarding experience at sea.

1. Plan to arrive at least 30 minutes early for a charter's scheduled departure time. This will provide time for vehicle

parking, check-in, carrying gear aboard, and taking sea-sickness preventive medicine, if needed.

2. Always check in with the group's divemaster before boarding the boat. If there is a sign-in roster, make sure it is signed upon boarding.

3. Do not leave the vessel's immediate dock area without permission after checking-in. Straggling divers are difficult to round-up on a busy dock and may delay departure time.

4. Consult the captain, crew or divemaster as to where to store dive gear and street clothes. Most charter craft have designated areas for placement of tanks, gear bags and other equipment for safety or environmental reasons. Never toss equipment in the first open space encountered.

5. Only street clothes and delicate camera equipment should be stowed below decks, and then only with the permission of the boat's captain. This prevents getting water on bunks and the crew's personal effects.

Because at times street clothes are not permitted inside cabin spaces, it is advisable to take aboard a large plastic garbage bag for wrapping clothes and towel in order to keep them dry while stowed in the diver's gear bag.

6. To prevent mix-ups or equipment loss, mark all diving gear before a charter trip for easy identification while on board. It is distressing to have to rummage through a jumbled pile of fins identical in design and not be able to locate the correct pair—usually just when a diving buddy is about to enter the water.

7. Before and immediately after each dive locate all diving gear and place each item in the gear bag. Keep bags away from walk areas.

8. Use the buddy system for suiting up, away from aft-center of the boat because this area is generally used by fully geared divers for moving to the entry point. Step *into* the weight belt when donning rather than attempting to whip it into position. Another diver might be in the im-

mediate area of the swinging belt and could be injured by its weight.

9. Fins should be put on last, and then only when the diver is about to make a water entry. Attempting to walk with fins across a heaving or rolling boat deck is dangerous.

10. Never enter the water until the boat captain or divemaster gives permission to do so. This is important for group control and for logging dive times.

11. After returning aboard from the dive, move away from aft-center of the boat and use the buddy system for unsuiting. Slide the weight belt down and step out of it, preventing the weight belt from crashing to the deck. Doff the tank and secure it in holding racks provided. Stow all other equipment in the gear bag as each item is taken off.

12. If sea-sickness occurs, stay topside and use the side rails when vomiting. Going below decks only intensifies motion sickness and will certainly provoke the captain's wrath. Sprawling on deck is also inappropriate because it will restrict passage on an already limited walk area.

As a final note on charter etiquette, remember that trash can be a very big problem on a small vessel. All charter boats have trash cans onboard; be considerate of the captain and the environment—don't litter.

PERSONAL GEAR INVENTORY

Any diver who anticipates serious participation in wreck diving should acquire those items of equipment that are considered minimum by experienced wreck divers. Although choosing the proper gear has already been discussed in this book, a listing of minimum equipment follows for review.

Buoyancy compensator. 35 pounds of lift, *minimum*.

Tanks. Double tanks preferred when water depth exceeds 60 feet or when wreck penetration is anticipated.

Regulator. Balanced first-stage type with submersible pressure gauge attached.

Emergency air source. Octopus second stage, pony tank system.

Wet suit. ¼" Farmer John style with bibbed hood, mitts and boots. For deep or cold water diving, a dry suit is recommended.

Diver's watch or *bottom timer*.

Depth gauge. Liquid filled with a back-up capillary depth gauge.

Compass

Knife or *diver's tool* and small back-up blade.

Light. With large reflector for maximum illumination. A small back-up light is recommended when considering penetration.

Decompression tables. Waterproofed.

Writing slate. With pencil.

Line reel. With high-strength nylon line for use during penetrations.

Collecting bag. Mesh type.

Accessories. Only as compatible with safety.

The "minimum" equipment used for safe wreck diving can become a staggering number of items, and packing all these pieces of gear can present a real challenge.

Most divers, regardless of experience, pay little attention to the fine art of packing a gear bag before a dive. Every diver has some method that seems to work well—until it is time to begin dressing. Then some amusing sights can often be seen on a rocking boat as various items of equipment are frantically yanked from the bag and deposited on deck around the diver's feet. With such a dazzling array of unorganized gear, it would take 15 minutes to decide which piece of equipment to put on first. Packing requires a systematic approach that will insure that nothing gets tossed on deck to be stepped upon or otherwise abused.

A good rule to remember when packing is: Last on—first in, first on—last in! By virtue of this rule, here is a suggested order for packing a gear bag.

Fins first, on bottom for rigid support; knife, line reel, mesh bag, followed by wet suit jacket; gloves, hood and boots, then wet suit pants on top. Scuba regulator, mask with snorkel, and light are last because they are delicate pieces of gear that must be on top.

When beginning to dress take only the item from the bag that is about to be put on. With systematic order, dressing even on a rocking boat can be made easier.

A lead weight belt should never be put in the gear bag. It must be carried separately to prevent smashing delicate gear and to prevent stressing the gear bag carry straps, seams and zipper.

As gear is being packed, the diver should think about the dive and all items needed for safety and comfort. Establishing a written equipment list and keeping it with the gear bag will insure that no items have been forgotten.

SUIT UP AND BEWARE

Suiting up on a rocking boat is one of the most difficult tasks for the diver to perform. Without help it is often impossible to syncronize movements required for dressing. Using one's buddy for assistance in donning equipment is prudent and will enhance individual as well as team safety. After both members of the buddy team have completely dressed, a final operational check of all gear worn should be performed.

A successful procedure has been developed by the author to provide a thorough equipment check just prior to water

Using the "buddy system" to suit up. The "beware" procedure should be employed to provide a through equipment check just prior to water entry.

entry. This system is known as "beware" and utilizes the letters B-WARE. It will insure that both members of the buddy team have properly functioning equipment, all items are correctly placed, and each diver is familiar with the operation of the other's gear.

All divers should go through a final B-WARE check before any dive regardless of how routine it may seem. No diver should ever believe that an individual's skills are above the need for a final evaluation of gear. A diver whose ego is so threatened by mundane pre-dive concern with equipment that the final check is ignored may be reduced to helpless fright when one item of life support gear is suddenly discovered to be missing or inoperative while underwater. An in-water emergency situation can quickly develop at any time while diving, and precious seconds can be wasted in attempting to discover how an unfamiliar piece of one's own or a buddy's gear operates.

The following steps outline the B-WARE procedure. Each step must be performed methodically, in the sequence indicated, to be certain that all items of equipment are checked completely and to prevent confusion.

The letters in B-WARE stand for:

B   Buoyancy compensator
W   Weight belt
A   Air supply
R   Releases
E   Everything is OK

Here is how the B-WARE check is used. Always stand facing the buddy so that there is an unobstructed view of all equipment. Proceed with the safety check, one buddy completely, then the other. *DO NOT B-WARE SIMUL-TANEOUSLY*!! It is important that each buddy vocalizes the check and not be embarrassed about stating the operation of each item as it is evaluated.

B   *Buoyancy compensator*

1. Ask your buddy if the CO_2 cartridge has been checked for puncture and that the detonator mechanism is operable.

2. Make sure that no tank harness straps interfere with BC inflation.

3. Check oral and auto inflation function.

4. Check overpressure exhaust valve.

W   *Weight belt*

1. Check that the buckle is right-hand release (opens by pulling latch to the right).

2. Check that buckle is dead center, directly over the diver's belly button.

3. Make sure that the belt is outside of all other gear.

Rule: The *weight belt* is the *last* piece of gear to be put *on*, so that, in all diving situations it will be the *first* taken *off*.

A   *Air supply*

1. Check that the air supply valve is fully on, then backed off one-eighth of a turn.

2. On reserve type valves, check that the reserve arm is up, or in the desired position, since some divers prefer that the reserve arm is down to start their dive.

3. Push the second-stage purge button to insure air flow.

4. Note the submersible pressure gauge reading and inform your buddy of how much air is shown.

5. Check that the gauge hose retainer strap is attached to the waist strap of the tank harness and not to the weight belt or BC harness.

R   *Releases, tank harness straps*

1. Check that shoulder releases are easy to reach and open.

2. Check all straps for twists; straps must run flat for comfort and positive releasing.

3. Check waist strap buckle for left-hand opening, and position it slightly off-center from the weight belt buckle to prevent confusing the two.

E   Everything is OK

Step back two paces so that there is a full head to toe
view of your buddy. Look over all gear placement once again.
If you are satisfied with the appearance of every item—then
give your buddy the OK sign while stating "everything is
OK"!

Using the B-WARE procedure is important for overall
safety before any dive is made. If an in-water emergency
develops, both divers in the buddy team will know how to
manipulate all items of equipment. B-WARE enhances se-
curity and comfort.

WATER ENTRIES

Several techniques are used for water entry from a boat.
However, because of a particular vessel's deck configuration, the choices may be limited. The best method of entry
is that which allows a diver the safest access in terms of
equipment control and the least possibility of personal injury.

Climbing down an access ladder provides the most secure
manner of water entry. With it, all gear will remain in place
and personal safety should not be compromised, especially
when a strong surface current is running, because there is
less chance of the diver being swept far astern of the boat.
Using an access ladder is the least glamorous method of
entry, but it remains absolutely the safest.

The most popular choice of entry techniques is probably
the stride-off entry. If this method is performed correctly it
should provide a secure means of getting into the water.
Stride-off entries can be made from nearly all deck areas
at moderate heights. When performing a stride-off from an
unusually high bow, the technique requires bringing the
legs and fins together during the drop so that the fin blades

The stride-off entry method which can be made from nearly all deck areas at moderate heights.

The back-drop entry method which almost always must be used from small boats.

absorb most of the shock upon impact with the water's surface.

Entries from small boats almost always dictate the use of a back-drop method. This is the least desirable entry but may be the only suitable means. The back-drop technique can be very disorienting and exposes the diver to a greater risk of personal injury.

All divers should investigate vessel configuration and discuss entry technique methods prior to suiting up. Choose the method that provides maximum personal safety and equipment control.

DESCENDING AND THE BUDDY SYSTEM

As soon as both divers in the buddy pair have made a safe water entry they should get together and snorkel swim on the surface to the boat's anchor line. At the anchor line they can exchange final pre-dive information, then signal descent.

Descents are best made feet first. This method will permit easy dumping of air from the buoyancy compensator since the diver is in a verticle attitude, thus keeping the air dump valve in its highest position. Descending feet first prevents losing sight of a buddy and will allow both divers to maintain contact with the anchor line for control. Ear equalization is also easier to achieve when the diver's body is upright, since this attitude sustains uncongested blood flow to the middle and inner ear.

Hand contact should be maintained with the boat's anchor line throughout descent. The anchor line provides a positive means of stopping descent for equalization, reduces the possibility of being swept down-current from the site, and assures physical orientation to the wreck once on the bottom. With the possible exception of clear, shallow water dives, it is always safer to use the anchor line for both descents and ascents.

DIRECTION OF TRAVEL AND NAVIGATION

Once on the bottom, direction of travel is determined by the lay of the wreck, environmental factors, or both. If a wreck is being visited for the first time, it is best to survey the site from a vantage point near the dive boat's anchor. This will provide the dive team with an opportunity to check current direction and strength, look for obvious hazards and record compass bearings for orientation.

Checking current direction can be accomplished in several ways. When diving from a boat, the anchor line will travel upward at an angle *with* the current because the bow of the boat swings against or into the current (unless very strong winds are blowing, in which case the dive probably will be canceled). When not diving from a boat, or when the divers have moved out of visual sight of the anchor line, current direction can be determined by observing the upward drift of exhaust bubbles or by purposely stirring up a small amount of bottom sediment and watching the flow of silt particles.

Observing marine organisms will sometimes indicate current direction. Many fish species will maneuver their bodies to point into the current for the purpose of feeding on drifting algae or plankton. Also, freely drifting jellyfish, algae or plankton blooms will provide the keen observer with current information.

When a current is detected on a wreck site, it is usually best to begin the dive by moving against the flow for two reasons. First, the divers have more energy at the beginning of the dive for swimming against the current; then when it is time to return, the dive team can use the current to drift gently back to the anchor line or exit point. Second, swimming into the current insures the advantage of superior visibility. If the wreck is lying perpendicular to current movement, the dive team can use the deflective effect provided by hull members for escaping its flow.

On wreck sites where the visibility is greater than 30 feet, navigation is usually accomplished by visual sighting in the direction of exploration *away* from the dive boat's anchor, or from the first-sighted reference if swimming to the area from shore. Good visibility decreases the need for accurate compass navigation on the wreck site proper. However, visual references should still be noted in order to provide accurate details of the wreck.

When diving on wreck sites with limited visibility, compass navigation becomes an important aid in preventing disorientation. Especially on large wrecks or structures with indistinct, tangled masses, the dive team may suddenly become confused about its position relative to the dive boat's anchor or other starting point. Disorientation on a wreck may produce anxiety, increased air consumption and panic. Divers must be aware of their position at all times during the dive.

At the beginning of exploration, a compass bearing in the intended direction of travel should be taken. This heading is noted on a slate along with the reciprocal bearing (180° opposite direction) for use when returning at the end of the dive. From the starting point, the team will swim to some distinct object that provides a visual reference within easy sight of the original position.

Each time the team moves to a different location on the wreck, a new compass bearing is taken and recorded on the slate before changing position. Using visual references in conjunction with compass bearings is referred to as pilotage. This navigational technique provides a positive means of establishing an accurate position at any time during the dive.

Compass pilotage for wreck diving is not difficult providing the navigator incorporates certain basic concepts of underwater navigation to insure precision. Because of the possibility of magnetic deviation from metallic articles on a wreck, the navigator should take compass bearings while

Diver checking compass bearing, providing a definite means of establishing an accurate position at any time during the dive.

hovering several feet above the site and note all depth changes when moving between visual markers. Remembering depth changes is especially important if the dive team is swimming to a wreck site from a shore position.

If a large obstruction is encountered in the direct path of an established heading, it will usually be possible to swim up and over the obstacle while maintaining the desired compass bearing. Swimming around an obstruction requires the navigator to keep track of the distance and compass heading adjustments needed to get clear of the object and return to the original course.

Determining when to return to the dive boat's anchor or to shore is based on remaining air supply. For most wreck diving situations the return trip should be started before the air supply is one-half consumed. Since the return course is generally a more direct route, because it is *with* existing current flow, one-half the air supply should be adequate. The returning compass bearing should be the reciprocal of the original starting heading.

SWIM TECHNIQUE AND BUOYANCY

As with other forms of recreational diving, the main objective in wreck diving is usually exploration. Many specific goals arise as a result of or are secondary to the initial exploratory effort. Certain swim techniques will enhance wreck site investigation through purposeful maneuvering and buoyancy control.

After years of lying on the bottom most wrecks acquire a great amount of silt deposits because water movement in the area is deflected and prevents the sweeping action normally associated with currents or tidal flows. Oxidation of metals, decaying timber debris and the effects of marine organisms contribute to turbidity when structural members of a wreck was disturbed.

To prevent ink-like clouds of silt from interfering with visibility, fin kicks should be modified and hand and body movements slowed. The traditional flutter kick is used primarily to swim over a wreck when surveying the site for salvage, penetration access, photography underwater or hunting. When swimming on or near the bottom a frog-like kick is a more practical method of locomotion because it allows the fin blades to move in a horizontal, side-to-side attitude rather than the up-and-down action produced by flutter kicking. Since the fin blades are sculling from side to side, there is less chance of stirring up bottom sediment. The frog kick is easy to master, efficient and restful. Minimizing the use of fin action maximizes visibility.

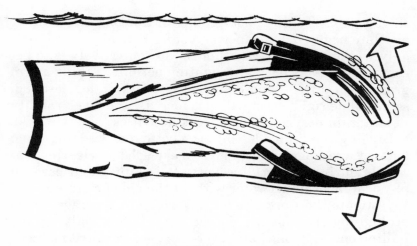

Flutter kick.

Once an area of the wreck is chosen for exploration, the diver should maneuver over the location by changing buoyancy. This is best achieved through decreasing lung volume. If the diver has fine-tuned neutral buoyancy, a continuous, slow exhalation will permit the body to drift gently into position.

The fins should be allowed to drop in such a manner that the tip of each blade makes almost imperceptible contact with sea bottom or superstructure. Easing the lower legs and knees down completes the technique of assuming a useful body position for survey, salvage or photography. If performed correctly, only a slight amount of silt will be disturbed, thereby assuring that whatever visibility exists on the site will not be destroyed.

After achieving the desired position on the site, the diver's buoyancy can be adjusted to the type of work to be accomplished. Wreck survey requires extensive maneuvering from one area to another, and neutral buoyancy should be maintained for easy position change during survey. When doing salvage work, taking photographs, or hunting underwater, absolute negative buoyancy is required. In other words, the only way to perform work underwater effectively is to become as heavy as normal weighting permits. Taking a firm grasp with a free hand on, or wrapping both legs around, some stable rigid object will also increase the diver's ability to maintain position during work activities.

Digging the fin blades into the bottom or wedging them against wreckage will add leverage needed for clearing materials during artifact recovery. It is difficult to perform physical labor underwater without the aid of rigid support. Investigate the work site for any source of leverage before attempting rigorous salvage activity. Finger-walking and pulling oneself along hand-over-hand are excellent methods of moving around wreck sites. Especially in potentially heavy silt areas such as passageways, deck houses, staterooms, or engine room spaces, these forms of mobility reduce the possibility of disturbing bottom sediment. The appropriate choice of swim technique or buoyancy is based on the object of the dive and existing silt conditions that affect visibility. Minimizing the use of fins and slow, purposeful movements should enhance visibility while working the site.

ASCENTS

When diving from a boat, the dive should end in close proximity to the anchor line. Returning to the anchor before ascending provides a secure feeling that safety is directly overhead. Also, any existing current flow on the wreck site dictates the use of the anchor line for ascents. Drifting downcurrent while underwater or upon reaching the surface can produce anxiety in the diver. Time spent exploring the wreck will have tapped the diver's energy significantly and there may not be enough endurance left for surface swimming against a current. Attempting to return to the dive boat up-current is an exhausting task.

As an established routine for diver safety, a charter boat crew will let out and tie off a buoyed trail line from the stern of the boat. This floating line, usually 100 feet in length, is placed so that when divers surface they can drift to it, grasp it, and wait in position for their turn to board the boat. Charter crews usually become alarmed when a diver is sighted on the surface out of reach of the floating trail line, because very few charter vessels have a chase boat, and as long as even one diver is still below they cannot pull up anchor to retrieve one who has drifted down-current.

If a strong current is running, an additional set of safety lines will be rigged from the bow anchor line along both sides of the boat, terminating where the floating trail line is secured to the stern. With the vessel prepared in this fashion, divers can ascend the anchor line, surface, then move aft to the trail line without losing hand contact with a safety rope.

When diving wrecks from a shore-based location, the dive plan should anticipate current direction from the time of leaving shore to returning. Accurate tidal and current flow is usually predictable during the brief period of a dive. An other important safety aspect to be considered is the direction of water movement and how the dive team can use flow patterns to their advantage.

The possible need to make a decompression stop before surfacing is another important reason for ascending the anchor line rather than making a free ascent. All dives should be planned in favor of no decompression, as mentioned previously. Nevertheless, the buddy team must be alert to the possibility that the dive profile may accidentally exceed time or depth limits and the consequent need for making a stop on the way to the surface.

Many divers, as a matter of routine, will make a safety

decompression stop following every wreck dive. This is done because cold water, deep dives, and/or hard work compound the problem of nitrogen absorption in the body. On a routine wreck dive additional ropes for decompression are generally not rigged. Thus, if a dive team is required to or chooses to make a decompression stop the only line available is the anchor line.

WATER EXITS

At the end of a dive, when buddy teams are ready to board the vessel, exits are performed by one diver at a time. In turn, each exiting diver moves from the trail line to the boarding ladder or platform. While holding onto a rung at water level the diver removes each fin, then slides a free hand between the heel strap and foot pocket so that one fin is retained over each wrist. The face mask and snorkel remain in place during exit, so that if the diver should slip on the ladder while boarding and fall back into the water a closed system is maintained, insuring that no water is swallowed and both fins will be available for use in regaining exit position.

In moderate to rough seas, the boarding ladder must be approached with caution. The pitch of the boat will cause the stern to rise, bringing the ladder high out of the water. As the stern settles the ladder comes down with enough force to severely injure a diver caught underneath. Boarding a pitching vessel requires accurate timing and close observance of the heaving stern.

Other divers waiting to exit should remain in place on the trail line and not move to the boarding ladder until the diver ahead is completely in the boat. During boarding, a diver is often very vulnerable to personal injury.

UNDRESS AND GEAR STOWAGE

After climbing aboard, the diver immediately moves away from the stern to keep the exit area clear. Each piece of gear can then be removed in an orderly manner that prevents heavy items from crashing onto the boat's deck. Members of the crew or divers already on board should be requested to assist with gear removal. As each article is taken off and disassembled it should be stowed at once, thus keeping the deck area from becoming cluttered with personal gear.

6

Hazards of Wreck Diving

MOTION SICKNESS

Motion sickness is one of the greatest problems the wreck diver must face and a vast number are susceptible to the effects of rolling seas. Some are so sensitive that they become nausiated even when the water is dead calm. Sea sickness can reduce even the most seasoned diver to total collapse. A palid, expressionless face is the first sign, followed by cold sweat, immobility and dry, heavy nausea. Once motion sickness sets in, the symptoms will continue as long as the diver remains on board the boat, and many are affected for several hours after returning to shore. One quote that appropriately describes the experience is "At first I thought I was going to die, then I wished I would!"

Many preventions and remedies are offered for the relief from motion sickness—watch the horizon, eat soda crackers, suck on lemons, get into the water—any of which may delay the onset of symptoms, but if the diver is susceptible, sea sickness eventually wins out. As with the common cold, modern medicine has been remarkably ineffectual in preventing or curing this age-old torment. Several over-the-counter medications are available which may be marginally effective, such as Dramamine, Marazine, and Meclizine. However, most divers report little or no significant difference between using these preparations and taking no medication at all.

Recent medical research and operational tests on Coast Guard, Navy, and Air Force personnel have revealed that the utilization of two common drugs taken in combination is so effective in the prevention of sea-sickness that the Coast Guard has approved each drug to be added to the ship's Medical Allowance List—promethazine (an antihistamine, commonly known as phenergan) and ephedrine (a decongestant). Because neither works by itself they must be taken in equal quantities. For maximum effectiveness they should be taken one to two hours before getting underway and thereafter at six-hour intervals as needed. Both promethazine and ephedrine require prescriptions. Therefore, a physician must be consulted in order to purchase these medications.

DIET AND LIQUID INTAKE

Because of its inherent complexity wreck diving can be a physically and emotionally demanding activity. No diver should underestimate the expenditure of energy required by the sport.

Even a seemingly innocuous boat ride to and from the wreck site may tap the diver's strength, especially in rough seas. Eating a high-carbohydrate meal prior to the days activity should be considered. Carbohydrates are necessary for sustaining energy over long periods of exertion and will aid in body heat production vital to combating the affects of cold water immersion.

Coffee, tea and cola-type drinks should be avoided prior to diving because of their diuretic effects, which promote dehydration by stimulating the kidneys to discharge body fluid. Retention of body fluids is important to the transfer of nitrogen during deep and/or lengthy dive profiles. Water, broths and soups are good choices for maintaining an adequate liquid supply in the body.

NITROGEN NARCOSIS

Divers are continually cautioned about the effects of nitrogen narcosis. This particular malady would seem to lurk in wait for any diver who ventures beyond 100 feet of depth. In reality, little is understood about how nitrogen acts to produce its intoxicating affect in deep diving situations. However, much is known of the symptoms of narcosis. Loss of judgment and skill, inability to concentrate, and euphoria are good examples of narcotic effect of nitrogen.

Diving physiologists have recently discovered that rapid descents, performing heavy work, swimming against strong currents or the effects of cold water immersion will increase the diver's susceptibility to nitrogen narcosis. This is thought to be caused by high carbon dioxide tension that acts as a catalyst on nitrogen. Further, it has been demonstrated that individual susceptibility varies with almost every dive, due to daily physiological changes.

Divers may learn to cope with narcosis at moderately deep depths, but overconfidence in one's ability to perform at depth will only increase the risks of deep diving. When participating in dives beyond 100 feet, the diver must be alert to any mental or physical change associated with the symptoms of nitrogen narcosis. Stopping activity and ascending several feet will usually return the diver to a safer condition.

COLD AND HYPOTHERMIA

The effects of cold water pose a serious threat to the diver that is often overlooked. In some climatic areas, being cold is expected and an accepted factor of any dive. However, when diving in water temperatures colder than 50 degrees the diver must be alert to the symptoms of excessive chilling and hypothermia.

In deep diving situations at cold temperatures, even a properly fitted wet suit is not enough to insulate the diver adequately. Deep diving causes compression of the wet suit's neoprene material, which results in a dramatic loss of thermal protection. At just 60 feet a one-quarter-inch thick wet suit will be reduced to approximately one-eighth of an inch due to surrounding water pressure. Deep diving, especially in cold water, therefore dictates the use of a dry suit to reduce the potential of hypothermia.

Hypothermia refers to the rapid loss of body core temperature due to immersion in cold water or air. Hypothermia in a diver is initially caused by paralysis of blood vessels in the skin, and it is of the utmost importance to maintain warm skin temperature through adequate exposure suit protection.

It is advisable to leave the water when shivering begins. The body is reacting to the effects of cold. Even one degree of core temperature loss will cause an increase in air consumption and impairment of motor skills or rational thinking.

CARBON DIOXIDE BUILD-UP

Retention of carbon dioxide in the diver's system is perhaps the most dangerous condition that can occur in diving. High concentrations of CO_2 have been shown to affect the onset of nitrogen narcosis and increase the incidence of decompression sickness. The problem of CO_2 cannot be over-stated and deserves continual self-observation by the diver during any underwater activity.

Everyone has at sometime experienced the feeling from overexertion which results in shortness of breath and fatigue. On land this presents little difficulty. But underwater, the problem of exertion is considerably more serious.

When shortness of breath or fatigue occur the diver may become overwhelmed with the feeling of not being able to get enough air through the regulator. This sensation of impending suffocation is very unpleasant and may produce panic.

A diver's ability to perform hard work underwater is naturally severely limited and many seemingly routine situations can lead to exceeding physiological limitations. Swimming against strong currents, prolonged heavy exertion in salvage, wasted efforts attempting to lift heavy objects to the surface, excessive cold or inadequate protection from cold, and increased breathing resistance in deep diving situations are all producers of high CO_2 levels and must be avoided.

Carbon dioxide build-up is easy to prevent if the diver monitors all physical activity and is constantly alert for signs of fatigue or labored breathing. Periodic resting with good lung ventilation will keep body CO_2 levels to a minimum.

DECOMPRESSION SICKNESS

The prevention of decompression sickness is extremely important for any dive performed deeper than 33 feet, and all experienced divers know they must closely monitor depths and times while strictly adhering to the guidelines contained in the United States Navy decompression tables.

Several other factors which must also be considered with

respect to decompression sickness prevention are age, fatigue, obesity, old injuries, cold water, alcohol or drug consumption, and rapid ascents that can pre-dispose the diver to decompression sickness even while staying within the limits of the Navy dive tables. Sport divers tend to display many of these conditions and should be guardedly conservative in their diving activities.

Two discretionary procedures in use by active wreck divers to avoid the problem of decompression sickness are planning a safety factor into the dive time–depth profile, and making a 10-foot decompression stop following any deep and/or cold no-decompression dive. In planning a safety factor the dive team plots the dive to the next greater time or next greater depth than actually occurs. The dive is thus pre-planned in favor of no-decompression and the profile strictly observed.

Stopping to decompress for five minutes at 10 feet of depth is a popular alternative to using the next greater safety factor because many divers feel this method is less restrictive on time available for no-decompression dives. However, one negative aspect of making the 10-foot stop would be the possibility of not having enough air supply to complete it.

When participating in deep dives with planned or unplanned decompression it is advisable to set up a special ascent line in anticipation of making a stop on the way to the surface. Such a line should be rigged in a manner that allows the rope to be buoyed at the surface. Buoying will ensure that a constant and accurate depth is maintained as the divers rest underwater during decompression.

Using a boat's anchor line or a line fixed to any part of a diving vessel is an inappropriate arrangement for a decompression line. A boat's rise and fall in rolling seas will create several feet of depth inaccuracy at a time when exact depth maintenance in decompression is the key to proper nitrogen elimination.

The buoyed ascent line is weighted to assure a good vertical attitude and scuba tanks, with octopus regulators attached, are tied off 10 feet from the surface to provide a precise depth for the decompression stop. This entire buoyed array should then be connected horizontally to the boat's stern platform to prevent divers from being cast adrift while decompressing.

Performing planned decompression dives is a serious undertaking. The risk factors involved are tremendously one-sided in compromising diver safety. However, there is a probability on any no-decompression dive that time–depth profiles may accidentally be exceeded. Usually this occurs because of distraction, entanglement or losing one's direction on a wreck.

If a required decompression stop is omitted during ascent, an emergenccy decompression procedure must be carried out. The "Emergency Procedure for Omitted Decompression" requires that immediately after surfacing the diver change tanks to ensure a full air supply and check the decompression time for the omitted ten-foot decompression stop. This must be done within three and one-half minutes of surfacing because nitrogen bubble formation occurs rapidly.

The procedure is performed as follows, based on the Standard Air Decompression Table with one minute between stops:

Descend to a depth of 40 feet.
At 40 feet, remain for one-fourth of the 10-foot stop time.
At 30 feet, remain for one-third of the 10-foot stop time.
At 20 feet, remain for one-half of the 10-foot stop time.
At 30 feet, remain for one-third of the 10-foot stop time.

Even though the procedure for omitted decompression is carried out, there is no guarantee that the divers will not get decompression sickness. It is advisable that the nearest recompression chamber facility be alerted to the possible need for treatment.

Symptoms of decompression sickness normally occur within one hour after surfacing. Especially in deep diving situations, all participants should be watched closely in this period for symptoms of bends. Divers should be instructed to report any unusual feelings during the post-dive time up to six hours after deep diving. Many severe bends cases have begun with only a slight itch or fleeting pain.

If decompression sickness is suspected be prepared to arrange transfer to the nearest treatment facility. Under no circumstances should in-water treatment be attempted.

The nearest facility can be located through the recently formed Diving Accident Network (DAN) co-ordinated at the Duke University Medical Center, Durham, North Carolina by calling their *DAN HOT LINE* (919) 684-8111.

In the United States there are seven regional hot lines that can respond to diving medical emergencies.

PENETRATION OF WRECKS

If there is a singular motivating force for wreck diving it must be eagerness to penetrate. A superficial excursion on any hulk leaves the diver with a feeling of unfilled adventure and a nagging desire to discover what lies hidden beneath those twisted plates or broken timbers. Still, penetration of any wreck can be a complex and hazardous quest. Impulsive attempts are discouraged because of the risk of entrapment. Any penetration plan must include special equipment required for and a knowledge of reaction to an emergency if one develops.

Special equipment for penetration includes a penetration safety line wound on a take-up reel, powerful dive light and small back up light, pony bottle, with regulator attached, to serve as an independent emergency air source, and an extra scuba tank with regulator attached to be positioned just outside of the penetration entry area.

Penetration begins by attaching the safety line to a sturdy piece of wreckage outside the entry point. As the divers maneuver more line can be let out as needed. Occasionally, the line should be single looped around some elevated object to prevent balling or tangling while moving inward. Such attachments offer better visibility and a greater chance of visual contact with the line as the divers begin retreat to the entry point. Hand contact with the safety line must be maintained at all times by all penetrating divers. The lead diver becomes the reel man while the buddy grasps the line in a loose-fist fashion. If contact with the line is momentarily lost a quick search overhead should reveal its position.

Heavy silting must be expected inside any wreck. Disturbing this silt will reduce visibility to near zero. Using the technique of finger walking, hand-over-hand pulling oneself along and fine tuning buoyancy will help prevent stirring up great clouds of silt.

At the planned time of return or distance of penetration is achieved, the divers follow the safety line back to the point of entry.

Experienced wreck divers use what is referred to as the "rule of thirds" for planning air consumption during penetration. In practice it means; one-third of the air supply is used for penetrating inward, one-third for retreat, and the final third for emergencies. The initial one-third of the air supply includes consumption during descent from the surface and any set-up time required prior to actual penetration.

If at any time a penetration becomes questionable in terms of safety the dive must be aborted. No adventure is worth compromising a diver's welfare.

Falling or shifting debris is always a hazard while penetrating. Physical contact through body movement against

The first step taken before penetrating the wreck is tying off the safety line.

All penetrating divers must maintain constant hand contact with the safety line while exploring the wreck interior.

wreckage should be minimized. Even exhaust bubbles must be monitored because they can exert significant upward force on overhead obstructions. If sounds of shifting objects are detected, it would be wise to retreat at once.

Entrapment can occur without warning. Close-quarter wedging results from attempting to swim through an area not large enough to allow the passage of diver and gear. Jagged edges or protruding pipes are potential entanglements with air hoses, harness straps and other diver-worn equipment.

Abandoned trawler nets, monofilament fishing line and ships rigging are constant dangers for even an alert diver. In most cases of entanglement the diver can work free without assistance if movements are slow and careful. However, the buddy should be informed of the trouble so that corrective action can be taken in the event self extraction is not achieved.

Wreck penetrations require a disciplined mind, proper equipment and thorough planning. The support of a well-trained buddy is helpful, but as a diver edges deeper into a mass of wreckage the realization of being truly alone becomes alarmingly clear.

7

Wreck Site Conservation

THE TAKING OF ARTIFACTS

There are only a few locations worldwide where the taking of wreck objects is strictly forbidden. Truk Lagoon in the Caroline Islands group of the South Pacific, and Tobermory, Ontario, Canada are two well-known areas of concentrated shipwrecks where sanctuaries have been established to protect the sites from the random destruction caused by artifact hunters. Viewing wrecks of these areas is an exhilerating experience. Because they are protected, large intact hull sections can be appreciated by all who visit for decades to come.

Shipwrecks in most areas, however, are seemingly fair game for salvage. The temptation of wrenching away a barnacle incrusted porthole or even picking up an old fire brick that once lined a boiler is difficult to resist. But, the taking

77

of artifacts can create complicated legal problems and is not a simple matter of finders-keepers as is generally believed.

An artifact, as defined by the United States Department of the Interior, is any object made or modified by man for his use, including, but not limited to petroglyps, pictographs (prehistoric and historic rock art), intaglios, rock alignments, paintings, pottery or ceramics, tools, implements, ornaments, jewelry, coins, fabrics, clothing, containers, ceremonial objects (items of religious or political significance), vessels, ship armaments, vehicles, structures (or remains thereof), and buildings.

This partial list covers a broad spectrum of objects considered to be artifacts, whether lost or abandoned, in the territorial waters of the United States. Further, the American Antiquities Act of 1906 protects any object that is at least 100 years old including, but not limited to, skeletal remains of humans or other vertebrate animals, extinct fossils, and objects associated with archeological, historic or palaeontological sites.

Removing objects from offshore wrecks by sport divers has been ignored by most United States authorities. It would be an impossible task to monitor the activities of the thousands of enthusiasts who participate in wreck diving, but shipwrecks of historical significance or newly discovered vessels suspected of carrying fortunes in treasure will certainly be watched closely by an assigned governmental agency.

Virtually every country in the world has laws governing the removal of artifacts. If there is any doubt concerning the legality of taking objects from a wreck, consult local, state or federal officials before salvage attempts are made.

The taking of artifacts from any wreck is discouraged, not only because of legal restrictions, but because it means the removal of a primary reason for diving a specific site. Each object attached to the ship when it sank increases the wrecks value in terms of diver exploration. Many once-gla-

Marking an underwater artifact site.

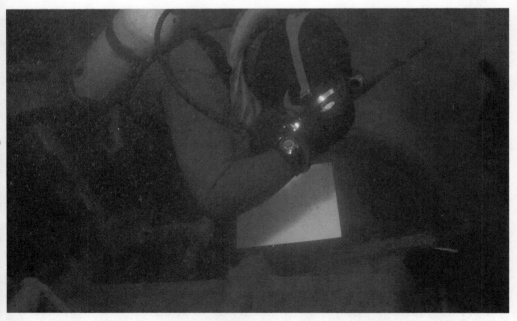

Recording collection data on a diver's slate.

Underwater photographer in full gear for recording details of an underwater wreck site.

Using a hacksaw to legally remove an artifact from a wreck.

mourous sites have lost their allure due to thoughtless artifact seekers who irresponsibly strip a wreck of its intrinsic
worth.

There is nothing more disappointing for divers who spend
long hours planning a first-time visit to a popular site than
to find only steel plates and sand. Self-denial of the urge to
remove an artifact will preserve the wreck for all divers to
enjoy.

8

Boat Terminology

The following terms refer to standard parts of any vessel, from small boats to large charter vessels, that may be used in wreck diving activity. The reader should study the drawing on page 46 to fully understand the specific meaning or location to which the terms apply.

AFT—toward the stern (back) of the boat.
AHEAD—in a forward direction.
ALOFT—above the deck of the boat.
ADMIDSHIPS—at or toward the center of the boat.
ASTERN—in or toward the back of the boat, opposite of ahead.
BEAM—the widest part of the boat's hull.
BELOW—areas beneath the main deck.
BILGE—the lowest point of a boat's inner hull.
BOW—the forwardmost (front) part of the boat.

BULKHEAD—any vertical partition (wall) separating compartments.

CABIN—a compartment for passengers or crew members.

CLEAT—a metal deck fitting for attaching and securing lines; usually anvil-shaped.

COMPARTMENT—living or storage area (room) below decks.

DECK—a permanent floor covering over any part of the boat's hull or compartment space.

DRAFT—the vertical distance from the waterline to the lowest part of the boat beneath the water.

FATHOMETER—electronic depth-indicating equipment.

FENDER—a rubber cushion, placed between a boat and a pier, or between boats, to prevent chafing damage.

FORWARD—toward the bow (front) of the boat.

GALLEY—the kitchen area of the boat.

GUNWALE—the upper edge of a boat's sides.

HATCH—an opening in the boat's deck.

HEAD—a marine toilet.

HULL—the main body of the boat.

KEEL—the centerline of a boat extending fore and aft which is considered the backbone of any vessel.

LADDER—any stairway on the boat.

LORAN—electronic position indication equipment (Long Range Aid to Navigation).

PORT SIDE—the left side of a boat, looking forward toward the bow.

RIGGING—the general term for all lines (rope) used on a vessel.

SCREW—the boat's propeller.

STARBOARD SIDE—the right side of a boat, looking forward toward the bow.

STERN—the aft (back) of the boat.

TOPSIDE—the top portion of the outer surface of a boat on each side above the waterline.

Index